Copyright ©2018 Flavio Davito All right reserved.

No part of this book may be used or reproduced in any manner whatsoever without written permission except in the case of brief quotations

Dedicated to my wife Brooke , who is the living example of unconditional love and to my children Finley and Tristan whose love for each other have been important to me.

INTRODUCTION

Does time exist? And is our future preordained by the past? And even more is the human brain pre-wired with an internal map?

Einstein on the Theory of relativity explained the possibility of an infinite and yet a finite universe: initially we imagine an Euclidean two dimensional space with flat rigid measuring rods but the whole sphere is a surface of constant curvature and under such condition and points and planes they traversed the whole spherical space and so it is easily to be seen that the three dimensional space is analogous to the two dimensional space and so the universe might be called a quasi-Euclidian universe.

Einstein Theory bring to life an universe that is now define in a geometrical way and where space and time became readable. Below the immensity of our sky our fellow vertebrates can navigate the passage of time easily as the neurons are able to fire high resolution images and anticipate events that will take place and coordinate time travel.

The four-dimensional space-time continuum and the Theory of relativity has grow out of electrodynamics and optics thus the invisible forces that navigate the universe are responsible for our lives and the environment that surround us like tiny invisibles strings released from the sunlight and attached to planet Earth.

Velocity as c plays a fundamental role in the Theory of relativity as a form which predict the effect produced on the light reaching us, the so called b-rays which emitted by radioactive substances consist of negative electrified particles that under the influence of the electric and magnetic fields explained the notion of these particles.

The method is simplified in nature as a series of events occurred at different places and where past, present, and future are equally real whereas our consciousness perceive time as an illusion as we shall see. Nature referred as plants and animals have their own circadian clock utilized as a light compass , plants for instance get energy provide from the conversion of ADP adenosine diphosphate to ATP adenosine triphosphate a process called photophosphory , the light rays are capture by the antenna complex and transferred to the photosystem , a complete reaction center, which contributes to sustained life to our planet.

In the same path is the structure of our brain which consist of billions of neurons communicating to each other through a system called synapses and just like a computer the neurons receive inputs and generate outputs, receiving signals and firing signals , and so the energy from the light affect and dominate our daily lives where the electrons or quanta jump from one brain to another and the cosmos.

Time and gravitational forces

The universe is not infinite any longer , the loops from the quanta of light are interconnected to each other like strings and holding the entire life of our planet together, but there is more , the electromagnetic fields carries radio waves, and transport the energy to the insects world in a spectacular way involving the visual display of the light production to a define passage of time between two biological clocks. The beetles , the fireflies , and lightening bugs among many others emit lights of different colors and frequency producing an enzyme called luciferase from the presence of an energy source of adenosine trisphosphate ADT and oxygen to produce carbon dioxide and light and able them to communicate with each other and some others are able to communicate by a sound emission , insects are able to recognize the passage of time , the Unified Field Theory is the link between two biological clocks .

The main components that allowed the interference between the light source and the insects stimulus are called photoreceptors a very complex nervous system of the eyes, where the light source change configuration of the visual pigments and triggering a signal as a chemical reaction or synapses into the brain and reading new gravitational fields.

But why are we not able to see those invisible forces ?

Our consciousness is separated by a different clock, Jean-Paul Sartre , a key figure on philosophy explained the significance detachment from the human consciousness , the reciprocal forces of revulsion which being and non-being exercise on each other , the evidence is the feeling of "anguish" or " nothingness" on our daily lives a form of transcendence state of outside itself, it is a state of a separate reality and human consciousness it is a sort of escape from itself, in short, the search for freedom .

Human consciousness thus is created from nothingness and consciousness is a reality attached with the feeling of anguish and fear of death but this identity can change in an instant with a feeling of pleasure , a constant interference between life and death or Eros and Thanatos , the fear of not finding yourself . We find ourselves in the presence of two different realities.

Freud did made a distinction between the "id" and the "ego" and so splitting the human psyche in two and making clear that we hold no position on the unconscious domain and according to

Paul-Sartre conclusion is that consciousness is here as a secondary phenomena a substitution of a different reality , the substitution is linked with the electromagnetic fields , those fields, are guided from the matrix of the universe. But according to Einstein , space and time is the same thing and so what are the components of the space that surround us ?

If the elements and substances form the universe and are able to read time can an atom keep track of time ?

The radioactive carbon produce by the sunlight determined the age of many scientific information and so life and death , those particles are dominated by neurons and carbons forming like a web that track the passage of time and according to the Theory of relativity gravity and acceleration warped into space allowing the quanta or electrons to jump from place to place forming the called " quantum leap' but before Einstein discovery Aristotle explained in the Metaphysics that movements and causes arise from the substance and Plato explained that the Forms are the cause of being and the Forms cannot became the causes unless there is a mover they were both right and wrong as we shall see.

Aristotle goes on and explained the mover as the essence or substance does exist as the invisible forces and it is not possible for an infinite to exist otherwise , infinite would not be infinite ., but what about the sensible substance that live in our minds ? we cannot link the mind with points , lines and shapes and so it would be impossible to have a science for both , the mind than belong to another science and indeed which one of these is the prior or master? We know the primary elements that constitutes our universe as fire, water, and earth but what about the sensible parts of our brain? And so there is nothing or something behind the human mind , something generated not in numbers and shapes but in love and passions thus the sensible parts of our brain are attached to some external forces that act as magnetic fields in short our mind is linked with an outer sources as we shall see.

Let's travel back to Einstein where explained that light could travel in a vacuum and it does not require a medium of transmission, it constant travel at the same speed 186,000 miles per second for everyone so time is relative to the observer thus everyone is subject to feel time differently.

He explained how the gravitational fields act as 'magnetic fields' which act like a stone that drop and produce motion to fall and receive acceleration , however the ray of lights are propagated curvilinearly in the gravitational fields bringing the example of the arbitrary curves draw on the surface of a marble table named as the Gaussian coordinated , the curves representing the rods that expand with the interference of the heat in the center of the table and holding a continuum of co-ordinates of the three and four dimensional shapes and creating the definition of space-time continuum and because the curves are always in motion , the law of nature, is influenced by the gravitational fields.

So the substances or quanta are always in motion and compare to a body of reference with a continuum expansion as the light produce heat but the contrary will happens as contraction, if cold will apply and doing so retaining a memory from the past so being an atom can keep track of the past and we are able to froze time and space . The typical example in nature is the snowflake which it is the representation of the invisible atom in the air and which instead of producing rainfall it produces snowfall. So the example A will bring a different result than example B. The A space and time is readable whereas the B is not but it is able to retain the frame of time because the frozen quanta is still living , the heat will able the universe to expand but the cold will be able to contract and retain time and doing so it is capable to reverse time .

We have to remember that the sensible atoms of our brain are attached to the magnetic fields of the cosmos and so our brain is not just attached to those external fields but also to the quanta that themselves are linked with the frozen past where the " not-being"of our consciousness lives daily and it reflects the cycle of the seasons.

The unified field Theory

But why the quanta are always in motion ?

Classic mechanics try to explain how bodies change with time imagining two clocks and Einstein bring the example as a man at the railway-carriage window is holding one clock , and a man on the footpath is holding the other. Each observer determines the position on his own reference-body but the embankment is fixed So here the example :

A   The carriage is in motion relative to the embankment.

B   The embankment is in motion relative to the carriage.

Upon the example the Earth will not move in uniformity in reference to those gravitational fields and so we will witness the space-time interval between points or events independent.

So in order to create an equation we have to use the co-ordinate K as the passenger K1 as the embankment K2 and as the train.

This event takes place in a three-dimensional continuum which according to Minkowsky consists of three reference bodies . Now we imaging the Gauss four dimensional co-ordinates of space continuum will change due to the such variables like heat temperature , the four co-ordinates will be x1,x2,x3,x4 and so the Lorentz transformation equation formula will be x1=x, x2=y, x3=z, x4=t or time .Hence, the coordinated will change shape with the heat temperature or expand and contract with the cold temperature but both will affect time and space as we shall see.

The reality of a multi-dimensional world is now defined with an equation corresponding to a rotation of the all the co-ordinates in the four - dimensional world. that is linked with the quanta of light.

But the lightning strikes before the sound, the speed of light is 186,000 miles per seconds whereas the sound speed is subject to variables such as temperatures.

We have to remember that the electromagnetic fields carries radio waves and vibrates within the insects' domain, the cicada for instance feel the vibration every seventeen years when they reach the hot gravitational field and feel the passage of time, the leap of time not perceivable to humans but visible to the ecosystem.

But there is more, under the sea, a common phenomena is the symbiosis ; the partnership between bacteria and the energy stored in chemical forms from the vent water to turn carbon dioxide into sugar, just like the light does in plants and so the tube worms under the sea are a bacterial farm completely independent from the photosynthesis but it reflects exactly what is happening in the universe The rays of light burn and produce carbon dioxide and it recycle back forming an endless cycle that sustained the entire life of our planet . The universe now looks like a gigantic web where the life of a spider is depending on eating is own web to survive.

The external environment such as the light and dark or temperature allow the circadian rhythm within a synchronized method to measure time, the biological clock of the insects allow the light waves to find orientation therefore assessing the passage of time like the light compass , finding direction and using the biological clock to find orientation from the movement of the sun.

But an experiment by Lindauer with honey bees trained to forage in the late afternoon and removing the hive overnight to a new location, he observed that despite the sun in a completely different angle, the bees were still able to communicate the direction and the distance of the food.

The circadian peacemaker , or oscillator, that control the rhythm is located in the brain , it is not an external receptor, and in beetles and crickets the peacemaker lies in the optic lobes as a communication center. .

The unified field Theory is the single theoretical framework that describes this interaction between these two biological clocks that interact each other, the light and the sound waves which are subject to change due to the variables from heat and cold.

Lightwaves and the soundwaves are both required to turn the gigantic clock and define time. But in the same time the light strikes before the sound so we have two different time and space recognition and they are readable only by the insects and invisible to us.

Galileo on the" Dialogues concerning two new sciences" explained the theory on the complexity of sounds as that one must observe that each pendulum has its own time of vibration so definite and determinate and that it is not possible to make it move with any other period given by nature.

Galileo is making an example of a musical chord : If we consider the sounds frequency descent along the arcs instead of the chords , all traversed in equal time . but these times are greater for the chord than for the arc , an effect that is remarkable because at first glance just the opposite to be true.

Since the two terminal motion are the same , but this is not the case as the shortest time and therefore the most rapid motion is that employed along the arc of which straight line is the chord.

Than knowing the numbers of vibrations which each pendulum makes in the given internal time one can determine the length of the string.

Than Galileo makes an equation according to the sounds of the strings: If my friend counts 20 vibrations of the long cord during the same time in which I count 240 of my string which is one cubit in length ; taking the square of the two numbers ,20 and 240, namely 400 and 57600, then, I say, the long string contains 57600 units of such length that my pendulum will contain 400 of them and since the length of my string is one cubit shall divide 57600 by 400 and thus obtain 144. Accordingly I shall call the length of the string 144 cubits.

Galileo formulated an equation for soundwaves which are now bodies of reference and those frequency/vibrations change with variables such as heat to match the vibrations of space and the change of time into another.

The speed of sound/string has a coil or shape which can be compressed or elongated , The more a string slows down , the more compressed it gets .The more a string speeds up, the less compressed and more elongated it gets and so we have a quantum cycle consisting of ; frequency, speed, compression, and length and the speed of the strings in space is different in different direction.

Galileo dialogue defined the ratio of the musical interval it is not determined by the length size or tension but rather by the ratio of the frequencies , that is, by the number of pulses of air waves which strikes the tympanum of the ear causing it also to vibrate with the same frequency.

Galileo makes the remark of speed versus sound/frequency ; repeating the stroke several times , now with greater now with less speed , the pitch was lower and higher and the marks made when the tones were higher were closed together but when the tones were deeper, they were farther apart and the speed increased towards the end of the sound became sharper and the streaks grew closer together, but always in such a way as to remain sharply defined and equidistant.

Velocity and light/sound frequencies becomes identical with the law of conservation of energy under which the Theory of relativity express the energy in the form of E=mc2 and mc2 is nothing else than the energy possessed by the body before it absorbed by the energy E, and velocity c plays a fundamental role in the Theory and in the same time light/sound frequencies are orchestrating our entire universe .

So our dimensions and time are directional proportional to the speed of the strings in space which are emitted in sounds and the speed of light recognize a different dimensional space and time, bring to light the evidence of two biological clocks that govern and sustain our planet including the life of our minds as we shall see.

The role of the coordinates and the string that attached the universe is simplified with an equation from Lorentz ; $x1=x, y1=y, z1=z, t1=t$ of an event with respect to K as the transmission of light or sound in a vacuum as the ray of light , and the sounds frequencies takes place according to this equation as the rays of light derived from this point of view.

Now the universe is not longer infinite , and so is time.

But what is the role of the human consciousness in reference to space and time ?

According to Mircea Eliade , humans recognizes two worlds , the profane and the sacred, the profane space experience is homogeneous and neutral and no world can come to birth in the chaos of homogeneity , the sacred space makes it possible to obtain a fixed point and hence orientation in the chaos of homogeneity . it appears and disappears in accordance with the need of the day .

And according to Paul Sartre ,the anxiety and disorientation that we feel everyday it belongs to the" nothingness ", the search for freedom, the door between the two worlds, the axis mundi, and life is not possible without the transcendence and so the sacred time is recoverable , it never changes or is exhausted , in short, it is a return of the original time known as freedom.

In order to investigate those comments we have to step back and take a look at the history of human mind and how consciousness had evolved.

The origin of the mind

According to the Greek philosophers the human mind developed in three parts: reason, emotion, and appetite, and started having a divine vision of the cosmos.

However, reason and sensation are direct opposites.

The search for pleasures and the avoidance of pain became the opposite truth of the modern Homo sapiens. Life does not depend any longer and completely by the existence of the soul and wisdom, but greed, lust, and desires become the supremacy of our modern thought.

Socrates explains: regardless that men are equal or not, they inevitably lead to moral corruption and regardless a democracy refuse to accept orders they will inevitably lead to corruption. To the one-man rule, "the man that tastes a single piece of flesh will become a wolf." And the process repeats over and over, like an endless cycle of political birth, decay, revolution, and renewal. The cycle will be transmitted to modern days.

The never-ending cycle of evolution and progress, from reason to emotion, from nationalization to globalization, from fictional economy to real economy, from regulated to unregulated laws, from generosity to greed. Greed for power, greed for love, greed for lust, which is an unstoppable machine. The definition in Latin is 'radix malorum est cupiditas", meaning "greed is the root of all evil."

Evolution is constantly changing and altering us; it is a never-ending process, equilibrium cannot be reached until..........

The cosmos mind of the genius

The genius mind is not created from nothing. They can breath, steal, and grasp the senseless motion, the particles, the anxious interconnected in the air, a space sensation that belongs to the ones that are capable of hearing and feel, a common language engaging the supernatural sensation and continuously braking barriers of the higher intuition creating the cosmos mind'. Geniuses breathe the same air; they live in the same moment where new ideas grow from previous ones, always in a progressive motion. Galileo help Newton, who help Mach, who help Einstein, in the renaissance evolution of the artists held da Vinci with the infusion of new laws of art, science and astronomy but their real gift was an extraordinary imagination. The Genius mind is attached to the higher mind of the Cosmos they realized that perception is always in motion and that every new discovery entails the loss of an old certainty, which was replaced by uncertainty, but the uncertainty is the place of " nothingness " They understood the dynamic of the senses, which nature and the cosmos possess. Everything in nature is subject to divine proportion. Proportion is not only found in numbers and measures but also in sounds and the wavelength of the air the real force of the cosmos and suddenly nature is transformed into geometry. Da Vinci's early experiments were aimed at determined how the incidents of light alter its appearance exploring the process of sight in his famous glass bowl picturing a man with a bent torso , his eyes reflecting across the glass reflected in the water and noticing a change of perception of the human eye. Leonardo spent decades studying the laws of optics. He conducted major experiments with the camera obscura, describing how the light, when it goes through a small hole in the camera, is inverted upside down. His discovery about sight and the parallel between the known and unknown came to represent both the pinnacle and the legacy of his life as a painter and researcher. The Mona Lisa was not painted in accordance with reality , it is a phenomena the Da Vinci perceived during his learning process and determined that light has different velocity that the rays of light are link within the retina of our eyes but those images are perceive differently according to the light and the dark.

Mozart symphony 25 in g minor is a product of a cosmic mind that introduced music to a totally new level. Mozart let the ' the voice ' running throughout his music and he transformed music forever. Galileo's ideas resulted from the observation of nature. He measured time the most primitive way: by the flow of water. Galileo could feel the phenomena of ' the voice ' from nature without the help of technical or scientific instrument observing things given from nature and adapted to a larger field and suddenly he was confronted by a new phenomena where he could read anticipated events .The process of the unconscious mind as an introduction of the cosmic knowledge already existing in the child like as a radar screen and ready to receive outer messages, his brain is preordained by the cosmos and he observe the world in a state of neutral perspective. With the intuitive knowledge of Da Vinci, Galileo, Mozart and Einstein allowed us to grasp a new universe never seen before. The harmonious formula decoded from nature where eventually Da Vinci represented with "the Vitruvian man", where all the parts of the human beings are in the same proportion in a new dimensional way, the gap between divinity and mortality as one. The new cosmos now is open.

But under our feet , the insects seem to already know the law of nature and every aspect of the insect life is virtually governed by a sense of survival and communication throughout chemical messages, called pheromones, which are produced by glands, and in order to be of any use, they have to be dispersed. Insects have developed a mechanism of spraying these chemical signals. This would not work unless there was an equally sensitive receiver to the other end. A female moth antenna tends to be thin and wire like, and the antennae of the male are usually bushy and function like radar screen used the screen the air of the sky. As little as one hundred molecules of pheromones in a millimeter of air are enough to start a male in pursuit of a chemical trail. Different populations of different species can have different blends of pheromone. For example, a female in Iowa may use a mixture of 97.3, but a female from New York may use a blend of 4: 96. Once mating is accomplished, they produce another chemical called an anti-aphrodisiac to prevent another male from courting them , and the male receiving discharge can remain in deadline coma for hours.

Why the answer is in the air?

The first school of philosopher was identified by three thinkers of the sixth century from Miletus: Thales, Anaximander, and Anaximenes, the primary study of these thinkers was about the functioning of the Cosmos, the relation between reality and appearance. Thales taught that everything derived from water as in Aristotle concept: from seeing, that the nutrient of all things contains moisture and the heat comes from it and it is sustained by it, and the seed of all things have moisture in nature and so water is the basis of moisture. Anaximander taught that the world and others world came to being out of the boundless and eventually will be absorbed back into it so it was an attempt to explain that a visible world was also an entity that produced other worlds somewhere resembling more of a law of nature. The third of the Miletians Anaximenes goes further to explain more the visible versus the invisible world and his substance , the air' which is surrounding the world just like the soul surrounded our body together and all the other substance are derived from the air , everything inside and outside our cosmos is based on everything we experience with air and its transformation and it is where the 'live substance' is the intensive factor with the human soul as an unchanging life .

The soul as a natural part of the universe and by which the living things and the world are held together. The thought was originated because the earth is finite and the sky as a physical entity with sources of energy that are moving and direct sources of the universe, those forces operate in everyday life and there is no creation from nothing or decay from nothing only change of substance. Heraclitus sees the continuation of the cosmos sustained by the differentiation of the opposite illustrating several paradox like hot and cold , wet and dry, living and dead which they pass from one state to another so are not conceivable opposite. And so everything in the world is produce by earth, air, fire and water but the air produce the unseen ' substance', there is no void but the infinitesimal crystalline matter in the air resemble the six sided particle of a snowflake. But then again earth, fire ,water, and air remain entangled by the cycle of the ' law of nature' where all the things arise by alteration out of the same thing, became different at different times, and return back to the same thing , this thesis confirm" the reverse of time ". Continuously rely on the movement and alteration, the cycle of life, death and resurrection, the ' symbol' of immortality ' after all a leaf fall from the tree but it is not dead. The cold produced contraction and solidification and the hot produce dissolution, motion and expansion, therefore hot and cold are active forces, while the wet and dry are acted upon them. So the first purest doctrine of philosophy tried to explain a rational knowledge of the ' void' and replacing it with the word ' atom' which became a revolutionary thought that eventually Einstein will establish new laws on his Theory of Relativity. The transformation of one body into another takes place when one of the two gives away to the opposite. So air becomes fire when water became moisture and earth becomes fire when the cold becomes heat. However equality does not exist since the majority of the universe consist of air and the only geometrical equilibrium could be obtain placing the earth at the center and all the other elements around it, three parts equally distributed water, fire and earth crating the 'law of nature ' and the endless cycle of the seasons. Anaximenes explained that our 'soul' which is air holds us together so the wind and air encompass the whole world comparing the breath of life to the air that hold animal and human together and this primary substance is the cause of its own motion. Even in modern days spirit and soul are molded together as one and as stated by Theophrastus " the air differs in rarity and in density as the nature of things is different; went very attenuated becomes fire, when more condensed, wind, than cloud, and when still more condensed, water and earth and stone; and all the other things are composed of these, motion is eternal the ' infinite ' but by this changes and cycles are produced. The air being the primary force of the elements it is also the primary source of the soul suggesting our 'immortality '. But as Spinoza explain that our body is finite once we compare to another body and our thought is limited once we have another thought however our body is not limited to our thought and our thought is not limited to our body but the body is eternal and our thought does not imply conception a cause is eternal to an inheritance factor and time is not apply. So with that being said, the body and mind think independently having separate chamber; 'a dualism'. Substance cannot be conceived from something external and so substances and elements are eternal. Otherwise, they will be limited to something that already exists and the universe is infinitive and so is time.

Reality versus cosmos perception

Our minds have the ability to convert pattern through a strong sense of intuition for their meanings. Just like in the transformation of the chemical animal behavior to the changes in the environment humans has the ability to transform the knowledge to intuition and intercept those outer "messages" outside our conceptual behavior. The existing world and the world given are two different entities; it is reality versus perception, the known versus the unknown, the seen versus the unseen. It is reality versus imagination where an intuition is represented by the "intuition knowledge," the process where our body is able to perceive without a mechanical and logical explanation.

Animals had been using this " cosmic knowledge" for millions of years , turtles can hatch in Florida and ride the stream to the North Atlantic gyre ,the clockwise current ,spending several years in the sea, leaving those current will be fatal and they could end up in the freezing water and die. However, turtles can sense the magnetic fields and capable of learning the position of these and use them as landmark to navigate and survive.

Images and sensation are interconnected with the eye. Color is an independent physical object that consider due our dependence of our retina, but it is also a psychological object and create different velocity. We obtain a different sensation once we look yourself at the mirror; we are stepping out from the domain of physic into a different domain.. Our vision is dealing with distortion, the inverted image of a familiar portrait turn upside down we do not recognize it anymore. Two retinas at work in synergy but not able to get the same sensation whereas the blind already possess a "cosmic world"the space sensation that are outside the physical and the psychological world.

So colors determine the velocity of light and dark just as the heat and cold determine the contraction and expansion of the Cosmos Humans have three types of color detectors that are tunes to different parts of the rainbow from red to blue. But the mantis shrimps for instance have ten within that range , plus another five or six that can see ultraviolet , to which we are blind but the eyes don't always need brainpower to get sensation even the eyeless sea urchin can sees the light sensitivity are spread across their surface and with their spines are able to break up the light into meaningfully" messages."

The inhibition of the conscious and the emergence of the cosmos mind

A conscious act is regards of a proof that it had been remembering as such, the hypnotized individual do not recall a past memory. Therefore, the definition simply is a condition regarded as memorable, a split personality, identified under different symptoms. One of this is the fixed idea, its function is to regulate the balance between the body and the mind and create equilibrium of our thoughts. A fixed idea is nothing more than an obsession to manage anxiety and fear once you lose control of your conscious. A disorderly condition may arise under the phenomena of hysteria due to which consequently, the mind triggers a variety of fantasies and dreamlike events and a more radical condition is the hysterical anesthesia (total insensitivity of pain) which represent the total decay of the conscious system and the total arise of the unconscious mind so a complete division of our mind. (Wlliam Henry Myers). An hysteric individual has lost control of the conscious system, his field of vision is much reduced and most of their object is view directly in front of his eyes and most of the conscious field of vision is lost. In addition, loss of personalities in the conscious mind can be experience in the sleep walking of all kinds including somnambulism and in the ordinary dream, the dreaming state is probably going on in the night but also in the day as a momentary brief lapses of attention including déjà vu experiences or fragments of images of past events unconscious related. In the dream we experience a world where is expecting us to enter a new realm and most of the time consist of conversations with another interlocutor . The somnambulist individual or sleep walker seems to know every single step of an action ready to be taken without the conscious help. In the extreme cases of hysteria, an individual could have several split personality.

But how can a single cell can hold together an independent and collective life ? A creature split into different persons with a separate existence yet related with the same brain, the same organism. The echinoderms like the starfish for instance do not have a centralized brain but they can handle several functions at once but the man has different bodies that needs to be nourished every day because desires and appetites change and transform the body and mind so desire and pain derive from one part of the body and not the whole and it is regarded as nothing because the need is taking care of the present and not the future. But the mind affected by the reason should considered the past and future equally so the mind should not seek a lesser pleasure from the present for a greater pleasure from the future .

Eros and Thanatos

Why do we always think?

Freud divided the mental process into pleasure and pain so the mental activities shrinks from any event might spoil pleasure due to the overall aim to return an organism to a pre living state of death. Therefore, the ultimate aim in life is death and the sexual self-preservation instincts fit this theory because their function is allowing individual to ultimately avoid or postpone death. Freud now formulated the conflict present in the mind: the self-preservation under the concept of Eros, the life/love drive contrasted with death of Thanatos so the Eros instinct fosters social actions created by Thanatos self-destructive behavior towards others as aggression and violence. This interaction could explain the behavior of both individual and society the idea of Thanatos drive; self-destructive tendencies were diverted towards other people, similar to how ruler might redirect a revolutionary impulse in his country by bringing about a war with another country. Freud's was interested in the fundamental tension between the individual and the society. He was a paradox in the fact that human created civilization to protect themselves from unhappiness. This is because all human are controlled by life and death drives.

He felt that civilization was vulnerable to radical disruption and reduced to unresolved conflict. He had already seen many violent social crises in his own country and in Europe during his lifetime. In his theories, he emphasized how participation in mass society releases deep aggressive impulses. Crises in society reveal aspects of human nature that are hidden in normal days of living.

Our body however is at motion or at rest, the mind is alert of a different body and so they change with time and so are the desires and pleasures and the mind has no knowledge of the existence of the body only his modification this ' duality' is carried thought the entire human life. The mind has a confused memory of his development and the body has the knowledge only after his modification so the mind does perceive the external body of his modification and perceive the existence of several bodies.

The body remain finite and perishable and the mind is develop by ideas in constant motion and so is built up by ' false idea' or opinions that change to the infinity and so the mind has a knowledge of second kind.

The mind also is affected by the external world and stimulus associated with it which seems to be outside the ' law of nature' thus the man rather than follow nature goes against it thinking of having complete control of his actions ., envy, anger and hate does not belong to nature ; the law of nature has a definite cause and the man is rather affected in many ways and his power of desire can increase or dismissed determine the mind to be active or passive but the body does not determine the mind but because of his nature of restrain violence and desires to satisfy the emotion, man will never be able to obtain freedom. Ultimately only the mind can create progress and think independently so the pain and pleasure are balance for the well being and the mind has the power to remove the existence of the body as the body can remove the existence of the mind, it is a never ending battle and love and hate is nothing more of the balance between body and mind Eros and Thanatos .

Love wants to retain the present and hate to destroy; the mind is in constant battle and creates other avenue like hope.

Time and the cosmos mind

Nature and earth, with the help of the organic world, give us traces from the past. However, what about our inorganic world where everything is deterred linked be the moment? Yesterday and today is only connected by an intellectual bound where time is always present. Time is hardly noticed and passed rapidly. From the moment that our conscious goes to sleep, our unconscious world is awakened. Time travels to a separate chamber, and we again experience a dualism with two separate intelligence and two different clocks. When you leave your pet for a while, it will greet you with much enthusiasm the next time you see it that it will seems that time does not exist for it. Humans can perceive a form of déjà vu, where an event is recalled from the past but we do not intellectually remember it happening. The memory from the brain's neuron reflects the one from a frozen molecule of the quanta which retain a single fact encapsulated in a box but it can actually release it within a specific temperature for future events, just like the animals use their reflex movement to hunt or foresee future events.

According to the quanta of light and the law of thermodynamics , the cosmos does retain energy and release and in doing so orchestrate the cycle of life which include time and the reverse of time.

As Einstein explained: the speed of light is our time clock once we challenge it. We can witness two different clocks. Suppose the horses come out of the gate at the exact moment and the course clock strikes noon. The horse I am on is traveling at the speed of light. In one second, I move 186,000 miles, but so does the light beam. When I look back, to me the clock has not change, time is frozen. For the others riders, the clock is moving and the clock seems normal. Normal now becomes relative because the faster we approach the speed of light, the more we find ourselves in our space and time.

As Mach explained : In order of stimulation from the strikes of a bell, we can only remember the last one, from the letters of the alphabet, for instance ABCD, we can only remember CD and forward, but we don't recall the letters backwards.

And ultimately the man is in prey of his own emotion and the desires change with age being in search for the self preservation of life , he is always in prey of emotion but those emotion like fear, greed, anger and power they all vanish with time. And ultimately man can destroy emotion or increasing the power of activity and altering an image from the past is from the present.

The imagination for the future is always more intense than the present so we exclude the present for a better future. The mind than think independently from the body, the concept of space-time is indefinite; it is the balance of body and mind 'the cosmic mind' that only matter.

Plato the first genius

In the republic, Plato discusses the metaphor of the cave, in which Socrates describes the world around us as a darkened cavern, across which images are projected in the form of a shadows. The people inside the cave had been born there and have been forged to watch the show since birth. If they were able to talk to one another, would they assume the shadows they saw was the real thing? They would believe that the shadows were the truth. But let's imagine that one of them escape to the outer world he may suddenly be dissolution and most probably would return the world of the shadows, but accustom to the new world with light, he would gradually begin to see planets, starts, moon and the sun itself and realizing the true light is the real world.

Plato was the first genius to foresee that the soul is like an eye; when it sees that on which truth is shining. The soul perceives and understands with his own intelligence and after the experience with Athens as a city-state with corruption; he felt that humans needed to understand their differences. Therefore, he developed the idea that the soul is separated in three parts: reason, emotion, and appetite. As such, he suggested leaving the cave to find the truth.

Democracy than must treat everyone equally otherwise corruption and civil disorder will rule the city indulging with pleasures and monarchy eventually prevail because the nature of man is as soon as taste desires he will becomes like a wolf and the process goes over and over creating the cycle of decay. The Roman Empire consisting of over 2.5 million square mile with over 60 million people overindulging the pleasures of money, power and greed vanished after 300 years.

The cosmos world of the honeybee

At the beginning, the bee as cleaning duties preparing combs and re diving eggs. Later the same bee will remove dead or dying bees from the hive. Then she becomes a nurse, secreting nutrients for growing bees. After few weeks they begum making beeswax and finally after the worker is ready to leave the hive and guard the hive from intruders and making sure that don't steal the honey, soldier bees defend the colony against intruders. Therefore, every bee has a specific task. The queen bee's job is laying eggs, and she can lie over a thousand eggs a day and the only contribution to society is inseminating or fertilizing the queen, which is actually lethal to the male. Queens are responsible for maintaining order in the hive and they are doing chemically producing pheromones that keep all the workers sexually suppressed when the population built up to excess numbers. Bees have extraordinary effective tactic against the predator they casting an intruder leaving the entire apparatus and causing the bee's death. But the stinger continues to inject venom into the enemy and the stinger left behind releases a pheromones , as she die a honey bee sends out a call for reinforcement and many more bees arrive at the scene to dispatch the predator. In the same time, they need to keep the same temperature on their home, when temperature drops, the bees form a cluster. The colder the temperature drop , the bees form a cluster and the temperature never falls below 17 degrees Celsius even when outside temperature falls to minus 30 degrees. (the Forest Unseen David George Haskell).

The superiority knowledge of the dream

Why dreams should mean something? In addition, why should they mean something different from their contents? After all, dream contents are very confusing and sometimes they make no sense. The main difference from the conscious thinking is that dream sets a reality with no beginning and no end almost like an endless movie full of fantasies. But as Jung clearly explained on his theories that conscious had been steady separated by the basic instincts, but those original basics instincts are not separated they just lost temporarily contact with our conscious and they are still trapped in our unconscious mind, dormant. This disassociation is the cause that man thinks is the master of his own soul but the reality is that here is a pre condition related with history and inheritance that we cannot deceive and is linked with the quanta of light from the cosmos In other worlds those experiences are still remain hidden in like a separate drawer and the unconscious knowing the existence ,create symbols in an archetype descendant .It worth to reflect that our unconscious is here to help . Dreams have an impeccable memory, which could trace the origin of our psychological time not by the days but by millennia and since primitive time believed in the unconscious being the father of the conscious and the conscious remain an entity that needs an ego to survive, the unconscious not so it freely adapts to travel spontaneously and have an historical path a seed already planted. Conscious impulses always have a relation with pleasure and unpleasant stimulus, which bring stability or instability in our mind. So pleasure is the dominant part of our impulses as there is always a string tendency to search pleasure and dislike pain, it is human nature. Children for instance that play with toys have already a sense of domination towards pleasure and he is obliged to repeat the same experience to keep him happy. However, with the passage of time the novelty factor of the pleasure diminished. Childhood and adulthood pleasures change and we witness the mind that forgot things that were pleasurable and became almost forgotten life experiences. So the pleasures that we experienced long time ago no longer exist strangely they vanished but they are replaced by other may be stronger desires as those pleasures do not satisfy us anymore , those pleasures and desires that were so important than are not important now .

Random thinking and the cosmos mind

Jung claimed that we have two ways of thinking: direct or logical thinking, and fantasy or random thinking. Out of the two, the latter is more spontaneous. Direct thinking causes fatigue to our intellect because it is not spontaneous. Direct and logical, it needs to be organized. Hence, in biological terms, we experience exhaustion. Originally, language was created spontaneously. During primitive times, imitation was used to express emotions. This imitation includes using the sound of nature to express fear, danger, and love. Typically, a cry of a monkey or the sound of a bird introduced a symbol, for instance a hieroglyph. Hence, our direct thinking is nothing more than a representation of the first sound of a bird or a cry of a monkey.

Speech is generated by a thought process, be it a memory recalling an abstract natural event, or sound the deaf or the mute already have their own language attached to a primitive word. If we do not think directly, thinking of the past and future simultaneously. Therefore, fantasy thinking is operated by pleasures and passion, a cosmos mind, intuition knowledge whereas direct thinking operates on reality. Much of the random thinking consists of many images atop one another. Much like a cartoon, an encyclopedia of thousands of images, that is the world of myths and symbols. The classical minds of the Greek mythology had tremendous vitality in describing myths. Everything was animated and conceived anthropomorphically. Primordial thinking is fundamental for today society where the use of technology as form of communication has resulted in the loss of our primary sensations, the memory of our ancestors, and their moral values. A lost civilization means losing the entire workload and effort that our ancestors had struggles to obtain for thousands of years for the sake of making an advanced society.

The significance of the celestial bodies

Living on earth, we are surrounded by a whole system of planets, all of which resolve around the sun. More interestingly, this "system" isn't just part of natural process that indicates the hours, days, and years that pass by, or lunar cycles that are completed on a monthly basis. This system is celestial masterpiece, which indicated each event that occurs on the planet earth. Although it may seem like random thought too many, this is a truth as it provides us with the answers to questions that have remained unresolved throughout the humankind history .If you give it more thought , you will realize that much like every other field that came into existence in the early years of human history, such as medicine , chemistry, and philosophy , astrology, and astronomy have both been right up there all along. The Mayan people were well versed with it as it's shows by the discoveries we had seen, and so were several other civilization that followed them the Egyptians and the Arabs. Astrology has long even a subject which humankind has utilized for answering several camped questions and make accurate prediction events and occurrences. Signs of the subject date back into the time of the Babylonians. It was found that Babylonian astrology was the very first form of organized astrology, which was calculated systematically. This is as old as the 2nd millennium BC. There is also speculation regarding Astrology being used in the Sumerian period , which indicate celestial ones day astrology have all even extracted from these ancient studies. It is safe today that Astrology has existed throughout humankind's existence, and all sort of civilization have sailed it to his or her use. This proves the significance of the subject and its importance to humankind. Discussing all this background regarding the subject is essential in prefer to prove to any non-believers who thinks Astrology holds no real meanings.

The cosmic thought an alternative dimension

The human mind is capable of things unknown. This proven by the fact that it possesses knowledge of things, which are way beyond our reach. Most often, we find you dreaming of perceptions of an entirely different reality. We are lost in a different world or living a different life. All of these dreams, thoughts, and occurrences such as the déjà vu , they all point to one thing and one thing only , which is the fact that human mind is capable of channeling and forming connections with outer sources which we are not even aware of and the truth is that each mind is connected with the individual that possess this mind is thinking about them, the human mind is not just connected with an outer source but are also connected within themselves.

It is actually due the fact that this mind is reaching out to an alternate dimension and picking up a drop of thought from the sea of thoughts. and brings further possibilities to light and if it can reach out the other dimension itself, it certainly means that there is no such thing as a new idea or concept everything in our ordinary lives had been preordained with an internal map and a specific software and the truth is that each idea has originally being thought before by someone back in history and ever since, it has become a part of the "sea" in the cosmos dimension. Now whenever every individual is dreaming of, or thinking about any thoughts, the truth is that their mind is channeling into the sea from the other world and thinking about it. This also proves that each human mind can be used as a channel for reaching out to this dimension or can be used to connect to I and help each other.. If we consider the power that the human mind possesses and follow this theory by introducing further possibilities, a completely new array of concept can be unlocked. For instance, is it necessary that only human being ' thoughts reside in this ocean of thoughts which is present in the other dimension? What if there are other entities thoughts, which present in the other dimension? What if there are other entities thoughts residing in this sea as well? Plants, animals, gods, living beings from other worlds or dimensions, the possibilities are limitless! While the other dimension may not be physical and perhaps entirely psychical, it certainly clarifies the déjà vu experience. Déjà vu can be exactly described as one of our mind's many travels out into this other dimension. Since there is an endless stream of thoughts, ideas, concepts, and emotions lurking around this realm, our mind can easily interact with any of them and experience them. This would explain how and why you feel said you have been throughout any given experience before exactly as how you are going through it right now. This is merely your mind reminiscing, feeling nostalgic, or simply recalling the last time it went throughout something similar. Since the sea of thoughts is endless, it is needless to say that many people before absolutely must have lived an experience exactly the way you are living it. As such, déjà vu is entirely explainable through this theory.

And the theory is linked with the quanta of light that ultimately self explained the Unified Field Theory .

The reflex movement

Animal sensation for instance has much greater accuracy as compare to human being. Their chemical process of perception is pure and reaches the realm outside the five senses so it is certainly true that in order to respond to an environmental stimulus, one does not required a brain. The more in isolation the intellect remain the more flexible is the behavior and so more finely attuned to the environment, consequently our instinct detect stimulus by a sensor neuron, this neuron deliver a message to an intermediary neuron, which connect, to a motor neuron and becomes a transverse orientation, a behavior in the animal world in which the body is aligned at fixed angles relative to the source of the stimulus. The sun, the moon, and the stars provide stimulus to which many organism serve as navigational guides for flying insects as well , hence memory is a physical event that leaves traces behind and where Mankind acquired its first knowledge from it.

The reflex movement is already present to a newborn baby and is set by stimulus quite mechanically .If he burns by a flame, he will remember in the future to avoid fire. This indicate that the self defense mechanism, as instinct, is already present from birth and a child would perish if was not able to suck.

An animal or insect can transform himself or herself into different coloration to adapt to particular circumstances as a self-defense from future enemies and the primordial senses as the preservation of the species is always in perpetual motion, it is a process that preserve the species when it is advantageous and destroy it when survival no longer needed so they retain a sense of memory and an inheritance process or DNA This act of anticipation for the future could also be a trace inherited from past species that left traces behind. The chloroplasts of the lichen for instance, remember seasonally to protect themselves with substantial energy to last over the rigid winter or otherwise they would die. It is a function of memory that has existed for billions of years . Freud studies attribute instincts to track a universal life back to his origin and restored by a memory related to an earlier stage of things. These things are the living entity that has been obliged to abandon under pressure of external forces. This result in the development of an "elastic force"; this new view changes dramatically the thought of an instinct as a factor that changes only with external stimulus. It is the contrary to an expression of a memory with a history of organic life. Examples of animal's life seems to confirm this view that instincts are historically determined so they would constantly repeat the same course of life. As a result, what it is left is the history of the earth we live and the relation to the sun as a form of life and origin linked with the quanta of light. An old state of things is an initial state to which the entity is striving to return. The state of death summarizes the aim of life and death, and is present in our daily phenomena of Eros and Thanatos. Instincts and behavior that have a memory or DNA attached to a mechanical force that is NOT changeable because they are pre-wired by an outer source and all instincts push towards a restoration of earlier state of things so instincts can be stimulated by external stimulus but also have a "memory" linked to a cosmic state. The salmon is another example , the young one hatch upstream in rivers but migrate down to the sea to finish growing , after spending many years in the sea they return to the same river to reproduce and start a brand new cycle ,and they return not just to any river but the " same " stream where they started, also different stream had different smell and the fish can memorized the smell of his own stream remembering the smell for many years ; those " messages" are always there in a dormant stage. A similar explanation can be found to the migration of flights of birds where there is a compulsion to repeat a phenomena of heredity and embryology

Time will not be far distant where mathematics, chemicals symbols and musical notes might be easy supplemented by a system of colors, signs, and symbols with phonetic alphabets

The internal dialogue

Freud felt that most of the ego was unconscious. The id was the primary source of energy, the primary components of personality. It is driven by the desire of pleasure and wants to obtain now. It is the only of the three components that is present at birth, our primary instinct. A baby who is hungry or unhappy will cry until his or her needs are met. There is no negotiation or recognition of limitation of what is possible, it is simply the Id demanding what it wants and needs. Once the child develops and interacts with the external world, they become aware, separate, and different for other humans. This is when the ego develops. As the child grow older, the ego becomes attached to the id which is been shape by the external world. The ego aims to try to fulfill the urges of the id. The costs and end fits action are considered before an urge is indulged, which is quite different from the id's wish to immediate grab what it wants regardless of the consequences. If necessary, the ego will delay gratification, allowing the fulfillment of the id's urge only at time and place that is appropriate . As Freud put it, the ego represents what may be called reason and common sense, in contrast to the id which contains the passions. If the urges of the id cannot be met, the ego also tries to discharge the tension that creates. The final element of the personality to develop is the super ego. It is an internal sense of right and wrong, which aims to perfect a person's behavior. Initially, parents provide children with moral standards; the super ego works to make the ego act in accordance with ideal standard so its function is a self-observation of conscience and of the ideal. Freud describes the id as the horse: the instinct. The ego is the rational driver of the chariot, able to guide the id, but never managing to gain complete control since the horse can overpower him. The superego is the father of the chariot driver, sitting beside him, pointing out where he is going wrong. Freud considered that the id could be inherited and that experiences of the ego, if repeated often enough and strongly enough by individual in successive generations could be transformed into experience of the id. Therefore, the id could harbor residues of the resistances of countless egos. Freud was fascinated by how these three aspects of the mind interact and come to conflict, with the ego trying to satisfy the id, without limitations imposed by the super ego, while also dealing with the external world. He described the ego as a poor creature owing service to three masters and so three dangers: the external world, the libido of the id and the severity of the superego. The ego is the strongest element so that it can manage the balance. If the id gets to strong, the person starts to act on their own urges without regards to anything else, while if the super ego becomes too strong, a person becomes controlled by rigid morals. Freud considered a balance between the id, the ego, and the superego to be the key to a healthy personality. The overuse of one of these elements could bring catastrophic consequences to the individual and society.

But the Ego is originated in the secondary stage of our life due to the fact that is linked to an outer source as the seed of our lives and it needs to be cultivated and educated accordingly .

The horsehair worm s life cycle begins when it hatches from an egg laid in a puddle or stream. He crawls around until a snail or other small insect eats him. Once inside the body, the larva wraps itself in a protective coat forming a cyst, then waits and becomes a worm. When it can grow no more, it releases a chemical that takes over the insect's brain. The chemical causes the insect to commit suicide and die in the water. The worm, with his already developed muscle, rips through the insect's body and goes free.

The activation of the second brain and the relation with symbols

Our unconscious level of knowledge "the cosmic mind" has an enormous repertoire of encyclopedia. It has its own cosmic intelligence and can trigger "messages and symbols" which is a process of oversimplification that arise from solving problem at our conscious level into the highest form of knowledge.

Imagine a computer with a huge encyclopedia that can reach its own occlusion after a random process and showing puzzle where in a matter of seconds create its own conclusion just like having a huge file cabinet with all the information from simple to complex and very complex.

The crickets absorb the art of listening at its best they used the so-called calling song characteristic by a particular frequency and pitch. Some species have regional dialects and accents and the females they can recognize the song of mates with great accuracy. The sound waves produce and transmit to the auditory nerve. The crickets are sensitive to a broad frequency beyond 20,000 cycles per seconds, which is beyond the sensitivity of humans, they can communicate to a mile distance; the male begins with a courtship song, and after the mating, is complete, males break into another song referred to as the triumphal song. In the same we should observe in the numeration were prior to our modern numbers Leonardo did Pisa alias Fibonacci lived around 1201 AD. He was able to decode nature through a sequence named Fibonacci numbers. He considers the growth of an idealized rabbit population assuming for instance a new pair, or one male, one female when they mate one pair every month and assuming they never die. The reproduction can be observed into the biological world and observing the rabbit sequence of reproduction he decoded mathematical sequence to higher level. The Fibonacci puzzle leads to a question of how many pair will be in one year. First month 1 pair Second month 2 pairs Third month 3 pairs and so on... So according to the sequence, we call it F = pair then sum of the numbers $F0$. $F1$. $F2$. $F3$. $F4$. $F5$. $F6$. $F7$. $F8$ 0. 1. 1. 2. 3. 5. 8. 13. 21 so the sum will be 2+1=3+2=5+3=8+5=13+8=21 ultimately nature decode mathematics. The same sequence can be seen in the biological world around us, such as the branching of a tree, the leaves of a stem, pineapples, a pine cone, a shell, a spiral etc. The mind is able to rebalance the body and mind quite immediately and able to open the " gate " to the Cosmos mind the gate of pure knowledge where messages are received for anticipated event, the Cosmos mind is the realm where time does not exist.

On a September full moon after sunset the corals of the Caribbean are covered with acne, a bundle of eggs glued together is soon released one colony after another. The eggs slowly arise to the surface creating fertilization and new baby corals are born after this precise" night" sex for the coral is over for years to come. The synchrony of time and space is inexplicable to anybody on earth this higher-level knowledge belongs to the Unified Field Theory.

Why do humans think differently

A state of consciousness is transitory; an idea conscious is no longer so a moment later that means that the thought is dormant but capable to become conscious, or we can say that it was conscious in a dormant stage but able to become conscious. Therefore, the reason why an idea or a thought cannot become conscious is opposition forces of moral resistance. Consequently, we can distinguish two identities with two mental processes that are in constant work and the ego supervises those activities. The present of the thought is always there intellectually, analyzing the moment in a continuous motion whereas in the animal world it is nonexistent and relies more in sensations of survival. The domain of the cosmos mind is the transparency of the thought, the oversimplification represented by ' Symbols'. In mythology Symbols are represented by experiences that are gather in our unconscious mind, they are facts that happened in primitive time and determined the origin of our mind and the origin of our civilization. Myths and Symbols are the encyclopedia of the Unified Field Theory, the future of our language which is equated to the universal language.

We can distinguish two languages: the unconscious thinking of the ancient time and the conscious thinking of the modern times. Myths than are not fantasies but facts that causally happened and belongs to our history and are perceived with our senses and in our dreams it is a cosmic language that the genius mind always listen and apply to his knowledge, the true divine force of our mind. The symbolic language determine who we are and that is why we all think different, our instinct have an history throughout a bloodline and ancestry; we do have the same common agreeable social behavior but our unconscious has an history on his own and it thinks independently, so individual can think alike, and his the Master of his own soul which brings meaning to his life in the universe. Dreams also is an interlocutor that have traces of our origin linked with the space and time of the cosmos, The Ancient Egyptian understood the importance of the dream as the access to the otherworld and it was known as rswt which means to be awake with the symbol of an eye, the Egyptian's priests were able to speak with their Gods and pay attention to the messages from an higher civilization. The Egyptians recognized the existence of another world and they spent 3, 000 years dedicating their life not to religion but only to death.

The supreme knowledge of the dream is our true interlocutor that is able to think in millennia whereas the conscious is only able to think in days; the geniuses recognized the importance of the domain of ' cosmic knowledge ' and paying close attention to the awakening stages and the stage where you fall asleep, the barrier between the conscious and unconscious world, two different worlds with two completely different time and space sensation.

A guinea pig without the labyrinth they can still react the same , the ducks without the brain removed at the contact of the water are still able to drink, removing the brain from a rat it can still function, the memory seems not to be the main hardware of our brain. Therefore, symbols and myths are the main driver of our thought and time is relative to it because it moves in constant motion. The sensory world of our body is detached from our mind and time is trying to pick up the speed but it is relative unless we live in the same space sensation, which is when it all comes together as one.

The primordial symbol

Our moral behavior also have a history of primitive time where instincts had been seeded into our mind so the presence and the complexity of out thought. This evolution takes place in various step from the primitive times or totemism to our modern time. Among these prehistoric organization or tribes, there were some who lived in a rudimentary stage where moral behavior was nonexistent. One so the social problem that primitives people had was incest's relations; a man found with a wife within her own clan was hunted down and killed by his clansman. This behavior took the form of prohibition of group incest and forbade marriage within the clan. The avoidance to look at each other became very strict and those instincts still exist dormant in our modern society. The Totem, as spiritual figure of an animal represented the first image of a father/god like figure in a form of absolute power and became the nucleus of social gathering and ceremonies. They became a ' taboo' (sacred), which is the oldest human unwritten code of laws and with time became traditions and finally laws. The taboo and the totem became the first sacred symbol referred to the origin of the earliest thing and the first form of pre religion establishing the formation of our moral behavior. As James Fraser referred to the tribes, in New Guinea, a tribesman who had been killed another may no goes near his wife and may not touch food with his fingers. He is fed by others, and only in certain kind of food, these observances could last until the new moon. Among the Monumbos of New Guinea anyone who killed in war becomes unclean and he may touch nobody, not even his wife and children. At some point, the primitive people feel the surge of a father/ruler figure and a chief is chosen from the tribe. In West Africa lived the priestly king Kikulu alone in the wood. He may not touch a woman nor leave his house, indeed may not even quit his chair, in which is obliged to sleep sitting. In West Africa when a king dies a family council is secretly held to determine his successor; the chosen is suddenly bound and throw in the fetish house where is kept until he consents to accept his crown. Meanwhile the savage of Timmes in Sierra Leone elected a king, reserves to themselves the right of beating him on the eve of his coronation, the act was considered a hearty goodwill gesture but sometimes the unhappy ruler does not survive his elevation to the throne. But with the Totem they start to recognize that an unseen world was actually existing , and the spiritual world of nature start to evolve into an anthropomorphic world and the animals and plants begin to have a

' soul ' a life never seen before . Shamanism became the first pre religion form of totemism and help to gather civilization around the world, now life start to have a meaning and a social behavior is sustain but evil forces were still there and bloodletting ritual became an evil machine and , idolatry start to emerge in various form and clans start fighting each other until an ever ending war until religion appears.. Primitive people start to recognize the first and most important element of life and meaning for their existence into the unknown universe, with the observation of nature, they decoded the cycle of birth, life, death and resurrection, the never ended cycle linked to the cosmos .

Cardioversion

Electromagnetic forces are spread all across the solar system. Due to these electromagnetic forces, all of the planets of the solar system are bound together as one unit, and this is how they function together, move about systematically. It is obvious that the human body is capable of producing current and has electrical current that flow inside and in some instances outside currents are needed to readjust the body back to its normal state. A clear example of this is Electrical cardio version, which is a method through which the heartbeat is restored back its regular rhythm or pattern by giving it brief electric shocks. This is why we observe the positive and negative terminal of magnets influenced each other, while terminals of two different batteries carrying the same charge repel one another. On both cases, the electrical charge flowing from these magnets influences the other and brings about a change, or electrical redistribution inside its counterpart. In simple words, each body carrying electric charge has its own electrostatic field, which can be influenced or altered by inducing another charge in its surrounding Every planet in our system can carrying his own charge A and a human being on planet earth is carrying his own B which is also charge of its own, then we can prove that both of these bodies, or at least the predominant body (which is A as the planet is the largest of the two) is well capable of influencing the other and bringing about a change in its electrostatic fields

Gravitational forces are the primary example of why people always staring to the ground instead of in the air, and so planets influence each other as well as earth and humans. Some of the position of the planets they are regarded as their stronger point while others at the weakest point forming geometrical angles with each others, influencing the charge from various points and positions giving a positive or negative influence to each particular individual called 'astrological influxes'.

Space sensation

(Einstein) Space and time cannot be absolutely defined, this is the crucial moment of the Theory of relativity which also confirm the relation between two biological clocks, one observe by us and the other by the insect's world ; there is an inseparable relation between time and signal velocity. Two events that appear to be simultaneous to one observer will not appear to be simultaneous to another observer who is moving rapidly and there is no way to declare that one observer is correct. Einstein said :Suppose that a lightning bolt strikes the train tracks embankment at two distant places, A and B. Realizing the need of definition one would require taking into account the speed of light 186,000 miles per second. His answer was: we would define the strikes simultaneously if we were standing exactly halfway between them but suppose that at the exact instant there is a passenger at midpoint M between A and B if the train is motionless the passenger would see the lightning strike at the same time but if the train is moving to the right the passenger will be rushing closer towards place B so he will see the strike at place B before seeing the strike at place A so there no way to say that events occurs simultaneously. This was a radical transformation because it means there is no absolute time, instead all the moving reference have their own relative time. The present does not exist as soon as you thinking about it, the present is already past, thoughts are related to the past, present and future so the "nowness" done not exist and physical time is just an illusion, the relation of time and space sensation as one which means body and mind as a whole is linked with the quanta of light as past, present and future as one. And each of us is linked within a specific channel to an hardware that is able to detect traces of the past, the present and predict future events. The invisible world now is the proof of a spiritual presence provided by the creation of the physical nature from birds to trees and stars, stretching beyond our solar system to the whole universe. From Galileo to Newton and then Einstein, the whole universe now is under investigation and the future of science will determine new laws of this complex universe. Now let us imagine a universe that is connected with other universe and the combination of them together forward the hypothesis of an endless time where things can be accomplished and appears in the same time but in different time zone. Quantum mechanics tried to explain this phenomenon.

A cartoon is made from a combination of several pictures put together to make a movie from a cine camera. At the same time, we are experiencing a simultaneous effect where three processes during a specific time will appear an anticipation of a specific event from specific time the Unified Field Theory break open the hypothesis of the new existence and development of six senses where intuition prevails and the anticipation of facts around us from our daily lives represent true reality. The empty spaces, the substance, the presence of other fields known as the chameleon particles (without which there will be no gravity) will be filled with lots of particles messages our magnetic forces and they became a way of communicating . Simultaneously the brain will be introduced in a new four-dimensional space and time a sort of electric waves where all the above mention senses are stored and where reason becomes a secondary instrument. A transformation will also be seen through humans social behavior where the need of writing or talking will becomes a secondary tool and phonetic imaging and messaging communication will take over a new dimension where everything will be moved disorderly and act randomly for our human mind

The ' substance ' or the unseen energy will be the next challenge for humankind . The collective energy of an ant colony can bring spectacular results , their primordial sense of survival and collective communication allowed them to see future events. Their body structures are far inferior of any other animal on earth, but an ant's colony outclasses anybody in any circumstances just with the help of their natural senses. Group living and communication certainly give them an advantage, using chemical signals called pheromones can exert their members of an immediate danger and capable of astounding feasts of engineering.

Charles Darwin formulated a theory that emphasized the variation rather than the type or essence. The transmutation of the species and the origin of new species , a process by which those individual conferring survival produce more offspring than those lacking such traits , was called natural selection . His biggest problem was neuter or sterile caste in insects society , which do no reproduced so cannot pass on beneficial evolutionary variations, the acme of difficulty the neutered ants faced not only from the fertile females and males , but from each other as well.

The Darwin theory had another problem , the mechanism of pangenesis. When Francis Galton transfused blood between white and black rabbits, he failed to obtain spotted bunnies as the offspring.

George Mandell pursued a number of studies of plants hybridization. He chose the Pisum Sativum (the garden pea) a plant that produced true breeding varieties that produced progeny from identical parents , every individual possessing two different factors for each trait, one inherent from each other, confirming cross breeding experiment that two different factors maintained integrity.

Thomas Hunt Morgan analyzed the fruit fly which completes a generation every ten days. He determined it was linked to one particular chromosome. The genetic processes that occurred on the fruit fly are identical to genetic process in higher organism. In the same species like bees , ants and wasps , males possess only a single sex chromosome which leads to a remarkable form of parthenogenesis or asexual reproduction.

And in the fertilized eggs of Hymenoptera turns into females , whereas unfertilized eggs turn into males . A virgin female, can therefore produce an endless line of sons, and in the case of the pharaoh ants , actually mate with one of those sons in order to produce females. Hymenoptera reproduction involves polyembryony , where at some point in life of the species, each cells of the embryo becomes autonomous and begins to develop on his own. (Bugs in the system, Berenbaum)

Artificial intelligence

In the new era of technology and computer machine we found a new interlocutor where our brain is related to it and so is our thinking, the brain likes to be addicted to new concept it is part of human nature but in doing so a new dimension emerge, and suddenly the symbols from dormant start to become more active and the artificial intelligence begin to appears, an inversion of common thinking and calculus known as isomorphism, which is the tool that translate from symbols to numbers and to images . Predominantly, we are back to the Genius way of thinking or even further to the time of the ancient Egypt and the hieroglyphics.

A definition of an event now becomes effective just with a single image so now we start thinking much more spontaneously. We do not need to create a sentence to describe something, it is a brand new encyclopedia, a new intelligence resembling the one from an alien civilization and it actually is, as we shall see.

Artificial intelligence gives the brain flexibility we can now gathering imaging in a geographical manner and pick them up at random simply without the manipulation of our rational thinking..

Humans are now capable to project themselves into the future as the billions of brain cells became a very sophisticated device attached with the universe, predominantly linked with a hardware that regulate out brain with a specific internal geographical map where ultimately it is already preordained by the past, in few words , the individuality of our brain had been already wired since childbirth from an outer source , an alien source, and it had been designed to write his own software .

But our fellow vertebrates are already able to navigate through space and time and doing so they feel the passage of time and create their own internal map without being able to be conscious and because of their senses are able to hear the space and feel the time .

But we have to remember that frozen time never change as the particles of the quanta can memorize time and return it back to his own mechanism hence, we are computer machine and we can only perceive the "now" as the present as our consciousness observe time at the moment But this concept is not acceptable because according to Einstein time is relative to the observer means individually we do understand clock time of an event that occurred at different places , in few words we do not understand time outside our skull and so the passage of time becomes an illusion but our fellow vertebrates are equipped with a time machine that are able to predicts the future and be engage with time travel.

It had been a long time since the Homo sapiens picked up the first symbol and a long progress had been made since he started using it exclusively for his evolution , but the first symbol was originated by the quanta of light that was recognized by the retina of our eye, the human eye is the most sophisticated detector of images and the retina is able to transfer those images to our brain cells emulating images and symbols known as isomorphism . But the retina of the eyes emulates different sensations it can reverse images ,distort them , and create different shades of lights and colors which are then transmitted into our brain which trigger a codification in random force. The motion and the transformation of images and symbols are connected with a cable into the quanta .

The cable from the insects eye are also installed within an hardware. The dragonfly's eye contains 20,000 ommatidia and is able to see things in a multidimensional way and transform the multi-dimensional world into one simple image ,a crystallization of an image ,and detect time travel where evolution is not existent. Our brain had evolved in such a way that we are now able to begin to understand time and be able to detect time travel backward and forthward as the billions of neurons are able to communicate to each other through millions of synapses , neurons receive inputs and outputs and are connected with others nut not just any others but collecting messages from a collective mind , our personal web or people that are family members that are close to us participating in a collective mind similar to an ants colony . Hence, the neurons that are transmitted from out brain cells are connected to a post and pre synaptic neurons which are able to provide electricity and fire those neurons creating a World Wide Web within us , this confirm that unconsciously our brain is able to anticipating future events and understand time .

The circadian clock never really change with time ,plants and animals are guided by specific magnetic fields and built a clock with the rhythm the ligh therefore the circadian clock needs to have the energy from the light in order to be sufficient and so it is sensitive to the UV radiation and benefit from it. The quanta of light take a fundamental role for life of a single -cell organism which included the ATP driven by the photosynthesis which in order to reach this form of energy , the quanta needs to go to the entire cycle from light to carbon and back to light again the light/dark cycle, and in doing so it becomes an oscillator of frequency .

In retrospective our consciousness is driven by external factors such as pleasures or pain , if we track the time with positive experience time does not exist but the contrary is with the experience from the feeling of pain, it is the transitional stage of the psychological domain, the realm of nothingness , the lap when we don't have control of our consciousness. It is psychological model manipulated by an external source as is not naturally linked with the law of nature, it is a distortion of reality because we are dealing with two different clocks .

The brain is able to produce stimulants to the neurocircuit and altering the notion of time and so those transmitters are influenced by the speed of the signal which have different stimulation effect but certainly with pain time becomes in slow motion .

We have to considered the illusion of time it belongs solely to humans , so the question remain is consciousness the tool that engage us to time travel ?

As we already understand light velocity is the foundation to read time and so is sound , the different intervals achieved by either music or light affect the neurons in the brain and those intervals have different velocity , this is the main reason that the level of stimulation changes according to the frequency of velocity from the light, thus the .pain and pleasure stimulus affected the stimulus from the quanta.

But there is more, the light and sound produce music and they are able to orchestrate an alphabet of unlimited notes of shades of light and sounds of music which influence the photoreceptor of our eyes which are able to read a grid like patterns of infinite ray of lights and so ,we are biological attached to the quanta and our brain is the oscillator with its own circuit and where the neurons are affected and responding at different state of stimulus.

We have to remember that the neurons are also affected by the radioactive carbon in a dormant stage and so have the ability to tell time from the past and foresee future events, it is the same formula from the law of thermodynamics where the entropy or a disorderly isolated system has a tendency to increase with the lap of time and so comparable to the frozen state of a snowflake , the atom can retain time and reverse it to its original state , with the help of the gravitational forces all the things in mass are brought forward from atoms to planets bringing mass and energy equivalent and so the gravitational forces are responsible for the construction of the universe , the force which causes two bodies attracted to each other described by Newton or the quantum gravity of everything described by Einstein in the Theory of relativity by the curvature of spacetime and this formula is encapsulated in a gigantic clock where billions of atoms are spinning backwards and in doing so are able to reverse time and maintained a state of memory , the clock behaves exactly like the hourglass which comprises two invisible bulbs connected vertically and the atoms are regulated from the upper to the lower one creating to measure the passage of time. The sky now looks exactly like the painting from Vincent Van Gogh, Starry Night, where billions of strings from the quanta are governing not just the stars and the moon but our planet including the flora, the fauna and us. Each one of us belongs to the gigantic mechanism of the universe, the unified field. We have to remember the lightening strikes before the sounds but both are required to make the clock works and so the quanta of light together with the sound makes the entire universe similar to an orchestra of infinite number of musical notes.

The Theory of relativity assumed this possibility as the past, the present and the future with the warping of spacetime permits to travel back and forth in time

Our circuit are wired with a purpose from our consciousness but our psychological domain is influenced by the self-preservation instincts of life and death which is the state of non-being and consequently consciousness is able to read time within different delays as the psychological domain has a separate chamber , a separate clock Under the sea , the salmon that return to the same stream that spawned after circumnavigate the ocean knows space and time as the salmon is able to see the "nowness ", the unified field that our consciousness fail to perceive. The unified field that is, subject to a common DNA in the animal world but in humans the DNA from unified field is subject to each individual , a personal DNA and at the end our brain is attached like a string to a gigantic hardware..

Therefore Minkowski spacetime is a combination of three-dimensional Euclidean space into a four-dimensional where spacetime interval between two events is independent and so if time vacuum exists in one reality is subject to exists in another confirming a multi-dimensional universe and so a multi-dimensional space -time.

As the human perception from the psychological domain has a very distinctive space sensation , the domain of deception is well known in human interaction employed within its sociocultural context as the human is not aware of his action while he is lying and affecting relationships and themselves , children already have the ability to deceive at young age within distinction between intentional and unintentional lying with the only purpose to benefit someone other than the self or to benefit the self without regard the effects on others or even to deceive to harm others . (Deception by Robert Mitchell).

Therefore humans are aware and have an intention to deceive only to protect the ego . It is a separate clock that we are not aware of it, an invisible fence that separate two biological worlds . But deception on animal has a completely different purpose , Tex Sordhal on evolutionary aspects of avian distraction display explained the theory of implication of deception for the purpose of parental behavior , Spending thousands of hours among the avocets described the aerial displays as two types , Dive bombing and diversionary , Dive bombing consisting of a swift flight at the predator this is the display used in mobbing ; it is an attack that intimidates the predator and the calls are often change in pitch at the point of closest approach. The Diversionary aerial displays consist of circling a predator with loud and persistent vocalizations accompany aerial display thus the intentional deception of the birds is strictly to preserve the nesting .

So why do we lie? the self preservation instincts from childhood to adulthood are exclusively to preserve life against death means winning at all cost but in the same time the same instincts don't recognize the space-time conundrum, hence, at this point we have to reconsider the reality of the question :Is this reality only for our consciousness?

For sure , at the end we are an alien civilization connected to another.

The Ba and the Ka

The ancient Egyptian rituals were based on immortality of the soul indicated with the mythological hybrid human- bird images and the relationship with the sky. The Egyptian sky was represented by the goddess Nut, and the Temple was represented as a maintenance of the Cosmos and formed a dialogue between humans and the gods ritual but they were not believers of any religion as well as participants of the church rituals The maintenance of their Gods was exclusively towards the Cosmos. Offerings took place, making sure of a divine connection with the gods, represented by the Ba, the spirit or like "flying in heaven likes a hawk or falcon." Later, with the cult, it became the physical form or the Ka. A similar ritual performance is noticeable with the pre-Colombian shamanism. The Temple remains as the house of gods were priests modeled a way of communication with the gods. Each morning, they assembled at the temple for duty with offering and nourishment for the gods. This is an important part of Egyptian life where everything surrounding their lifestyle was based on the daily offering and strongly believing in the immortality of the soul. Most of the food offerings consisted of flowers, food, grain, vegetable, and meat. Right after the ceremony, the priest performed "the bringing of the foot "which means grabbing a broom and sweeping the footprints backing out from the temple.. This meeting with the living and the dead reflected the Egyptian cycle of life, death, and rebirth with the purification ceremony not just in the Temple by also in local ceremonies like the Osiris's "beautiful feast of the valley" where Amen is removed from his sanctuary , placed on a ceremonial boat and carried through the Nile .

The procession then travels from East to West, symbolizing the transition from life to death or the passage of the sun- moon. Once the procession reached the heaven, it becomes the Akh, and all the sins were expelled and the soul became perfect again. It is important to notice that the mythological figure of the human hybrid animal is always present in the form of vulture, ibis, falcon, and winged cobras as the protectors and guardians they represent the transports to the heavens with the Ba as the messenger of the god/goddess. The Egyptians strongly believed in the divine senses of the Gods, and they could sense the presence of the gods through smell, sight, and intuition, just like a bird could sense it. The dead could communicate with the living and the living had access to the dead.

Beyond the infinity

Spinoza self explained: A thing is called finite of his kind when we found another thing so a body seeing another body is finite and the thought having another thought is finite. Everything exists either in itself or in something else and so the knowledge depends on something a cause so a thing that does not involve existence does not exist. In the universe does not exist two things of the same nature and substance cannot be produced by something external therefore exist from its nature and its nature therefore does imply existence, the existence can be finite and infinite but finite is limited to something else that already exist of the same kind so infinite exist. God as the attribute of infinity than exist. Every reason must be assigned either exist or not, being finite or infinite. If a triangle exists, a reason must be granted for its existence. A potentially of non-existence is a negative power and a potentially of existence is power so something that exist is more powerful that something that doesn't exist and that is absurd therefore nothing exist or else the infinite exist also God as infinite exists as well.

The civilized man and the savage man

Rousseau vividly explains the differences which are sort of inequality of human race, categorizing the first in a physical form. This use to the fact that it is established by nature, which means body, strengths, mind, and soul. The soul inequality consists of different privileges such as being more richer, more honored, more powerful than others so the question is those who command are superior to those who obey? In addition, his wisdom and intellectual strength found in the individual proportionally to their power or wealth? It follows by saying that no animal naturally makes war on man except the case of self-defense or from extreme danger, nor does an animal exhibit towards a man any of those violent antipathies, which seems to be the mark of a species destined by nature to serve as the food of another. Man has other enemies which are much more intimidating and against which he has no same means of defending himself, his natural infirmities, old age, and illness of every kind. A proof of our own weakness is that the savage man dies without others to noticing that they ceased to exists, and almost without noticing themselves. The extreme inequality of our senses. The over elaborated food of the riches, the bad food of the poor which they often go without, those late nights and excess of all kind, fatigue, exhaustion of the mind , the innumerable sorrows and anxieties that people an all classes suffer and by which the human soul is constantly tormented; these are the proofs that most of our ills are our own making and we might have avoided all of them if only we adhered to the simple , unchanging and solitary life that natures ordained us. Medicine may be good for us but a sick savage man who abandons himself has nothing to hope for except from his sicknesses, all of which makes his situation very often preferable to our own. Let us therefore beware of confusing a savage man with the man we have before our eyes ; the horse, the bull, the cat, the dog all have a more robust constitution , more vigorous , more strength and more spirit in the forest that under our roof. They lose half of their advantages on becoming domesticated and one may say that all our efforts to care for and feed these animals have only succeeded in making them degenerate. The same is true even on man on himself in becoming sociable and slave he grow feeble , timid, and servile, and his soft and effeminate ways of life complete his strengths and courage . Furthermore, the difference the difference between the savage man and domesticated man must be greater than the difference between wild and tame animals, for since the man and beast are treated equally by nature , all the commodities which can give himself is beyond these give to the animal he tames. Being naked, homeless, and deprived of all those useless things we believe so necessary is no great misfortune for these first man and above all no great obstacle for preservation. The savage man cannot but enjoy sleeping and he must sleep lightly like the animals, which think little and self-preservation is his only concern. His best-trained faculties must be those, which have their main object to attack and defend or subdue a prey or to avoid becoming the pry of another animal. On the other end , those organs which are developed only by soften and sensuality must remain in rudimental state, including any kind of delicacy in his five senses , he will have touch and taste in an extremely course form, but sight , hearing and smell in a most subtle form. On the moral aspect, I see all the animals only as an ingenious machine to which nature has given sense in order to keep itself in motion and protect them against everything that is likely to destroy or disturb. I see exactly the same thing in the human machine with the difference that while nature alone activates everything in the operation of the beast, man participates in his own actions in the capacity as a free agent. The beast chooses or rejects by instinct, man does so by an act of free will, which means that the beast cannot deviates from such rules to his own prejudice that is why a pigeon would die of hunger beside a fish filled with choices of meat, and a cat beside a pile of fruits grain. Man abandons himself to the excess that bring on fevers and deaths because the intellect deprives the senses and will continue to speak when nature is ' silent '. The soul of a savage man, which nothing disturbs, has only the sensation of its present existence. Without any idea of the future, his projects are limited like the horizons. If we assumed his mind to have all the intelligence and enlightenment it needs to have, what progress could human race make among the animals in the woods? I ask if anyone has ever heard of a savage man in a condition of freedom ever complaining about his life and killing himself. Nothing could be more miserable as a savage man tormented by passions and arguing about his life and killing himself, nothing could be more miserable as a savage man tormented by passions and arguing about a state different from his own. In instinct alone, man had all he needed for living in a state of nature in a cultivated reason. He had what is necessary only for living society; the calm of the passions and the ignorance of the vice prevent him to do evil. Imagination, which causes so much havoc among us never, speaks to the hearth of the savage. Nature responds to his needs and once the need is satisfied, all the desire is extinguished. At the end, the savage man wandering in the forest without speech, without home, without air, and without a relationship, he is equally without any need of his fellowmen and without any desire to hurt them. Being subject to so few passions and sufficient himself, he had only such feeling and knowledge as suited by his condition this inequality in the state of nature remains to explain its origin and progress in the human mind.

3001 Space odyssey

A computer machine is able to identify very complex problems, but above certain level of complexity, a computer ceased to be predictable and start to do things on his own account so a machine needs to be program in order to act . Ancient civilizations thought that Earth was the center of the entire universe. They believe that our planet was capable to sustaining life, but they also believe of the existence of another world, the otherworld, the afterlife and the immortality of the soul represented by Primordial Symbol of the pyramid which self explained the history of afterlife and the cycle of life, death, and rebirth. The ancient Egyptian perceived the existence of two different realities; two different time's zone and they dedicated their entire life to that. A model for the world from the most sophisticated civilizations ever exist.

The archetype of our civilization changed many times from the first ' voice ' of the hieroglyph to the Greek ' thumos' evident in the Iliad where the gods are always communicate with the humans to the sacred structure of temples and churches which transformed the primitive man to a modern communities . Than with progress and the beginning of the marketplace the ' voice ' started to diminished and the beginning of a new internal dialogue emerged only to satisfy the desires and humans appetite nothing else.

The universal language and the world of imagination that touched so many Geniuses like Da Vinci, Galileo, Mozart and Einstein is in a dormant stage but six million years ago, the Homo sapiens start to use the first tool then curiosity emerged and the invention appear at fast pace, humans are soaring to evolution but something went wrong and technology start to take over the first man tool. The Great Pyramid is still here to observe us to see as far can we go but all of a sudden, the void ' the substance' exists and a new brain with firing neurons seems working at high speed but humans struggle to walk again and the evolution is a total failure.

Man knew how to breathe air but computer does not and it makes mistakes. The computer won and the man start facing the uncertainty beyond the infinite. May be the last cycle of birth, life, death and resurrection is underway and may be the 'cosmos mind ' remains our only tool to listen. The Cosmic moments of Mozart and his symphony 25 in g minor is always there signify the universe and so the master Da Vinci with symbolic representation of the Mona Lisa .

People heard of Diogenes walking the street of Athens with a lantern, looking for an honest man and never finding one. Diogenes believes that man needs nothing in order to be happy. At some point Alexander, the Great heard of him and wanted to meet him. Alexander finally found him sunning outside the gymnasium and said: "I am Alexander the King" " I am Diogenes the cynic," replied the philosopher. " is there any favor that I may bestow upon you?" Alexander asked. Diogenes looked at him and said, "Yes, stand out of my light" Alexander following this event said that if he could not be himself, he wanted to be Diogenes. But no one asked Diogenes if he would prefer to be Alexander the Great.

Quantum Of Life

Lo frate Sole,Lo quale e' iorno, et allumini noi per lui.Et ellu e' bellu e radiante cum grande splendore.Per sora Luna e le Stelle. Il celu l'ai formate clarite et pretiose et belle.Per frate vento Et per aere et nubilo et sereno et onne tempo,Per lo quale , a le Tue creature dai sostentamento. Per sora Acqua.La quale e' molto utile et humble et pretiosa et casta.Per frate Focu Per lo quale ennallumini la nocte:Et ello e' bello et iocundo et robusto et forte.Per sora nostra madre Terra,La quale ne sustenta et governa,Et produce diversi fructi con coloriti fior et herba. Per sora nostra morte corporale,da la quale nullo homo vivente po' skappare Beati quelli ke 'l sosterranno in pace ka la morte secunda no' l farra' male.

www.ingramcontent.com/pod-product-compliance
Lightning Source LLC
Chambersburg PA
CBHW050059230526
45470CB00004B/1604